# HOW TO MEMORIZE FORMULAS IN MATHEMATICS

*BOOK-1*
*CALCULUS*

**Rajesh Sarswat**

*@ Rajesh Sarswat*

*First Impression: 2017*

No part of this book may be reproduced or utilized in any form or by any means, electronic or mechanical including photocopying, recording or by any information storage and retrieval system, without permission in writing from the author.

# TABLE OF CONTENTS

- HOW TO MEMORIZE ........................................................... 1
- FORMULAS IN ................................................................. 1
- MATHEMATICS ................................................................ 1
- Acknowledgments ............................................................ iii
- About the book ................................................................ v
- 1 ...................................................................................... 9
- HOW TO MEMORIZE FORMULAS ................................... 9
- 2 .................................................................................... 19
- TRICKS TO LEARN FORMULAS OF DERIVATIVES ..... 19
- 3 .................................................................................... 31
- TRICKS TO LEARN FORMULAS OF INTEGRATION ..... 31
- 4 .................................................................................... 49
- SILLY MISTAKES IN CALCULUS ................................... 49

# Best Seller in Pre-Calculus Category
# (www.amazon.com)

# Acknowledgments

I dedicate my fifth book to all my students for inspiring me to write such a book.

Rajesh Sarswat

# Next Book of the Series
# How to Memorize Formulas in Mathematics
# Book-2 Trigonometry

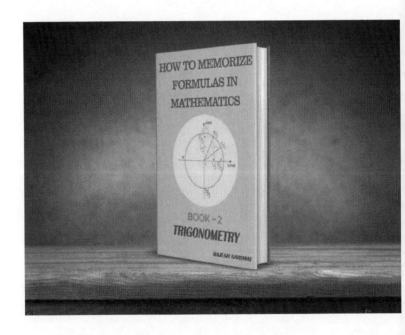

# Amazon bestseller in Trigonometry Category at www.amazon.com

# About the book

I can recall the day when my mathematics teacher taught me the formulas of trigonometry for the first time. I was in Class IX at that time. I had no clue, what was happening in the class. As soon as the chapter was over, I was restless and lost my sleep as I was unable to cope with a large number of formulas taught to me with no method to retain these results into my mind. Though I still believe that practice and learning derivations are the two best methods to learn all the formulas of mathematics, I always had a view that some alternative techniques or methods should be taught in the school to all the students to deal with such large number of formulas and their applications.

During my interactions with students over the last 25 years, they have indicated that the most terrible thing about math was learning the formulas. It can sometimes be very tricky to understand and remember the mathematical formulas since there are many for every concept and that too with various applications. If a formula is put wrong regardless of how small is the mistake, the whole problem goes wrong. Moreover, that inspired me to write this book.

With the tips given in this book, students will be able to memorize most of the formulas with better understanding. The book covers two aspects of formulas learning. One is general aspects, which apply to all branches of Mathematics. The book covers all these elements in detail in the first chapter. The second element is some particular techniques, observations, derivations which will help students to learn the formulas of a specific topic in Mathematics (Calculus, Trigonometry)

This book is still in the beta stage, and I will endeavor to keep it improving gradually, based on the suggestions and feedback received from my students and the readers. For any suggestions or any doubt about the topics covered in this book, readers may contact me at the addresses provided below. I will be glad to aid the readers to the best of my capacity.

**Facebook Pages:**

https://www.facebook.com/avoidsillymistakes/
https://www.facebook.com/quickcalculations

Email: rsarswat.rs@gmail

# BY THE SAME AUTHOR

You would also like to read another book **(a bestseller at www.amazon.com)** authored by me titled **"BE A HUMAN CALCULATOR."** The book is about observation based calculation tricks for improving the calculation speed of the students. The book will not only help the students to do calculations at a faster rate but will also assist them in reducing their computational errors by improving their interest and creativity in mathematics.

# 1
# HOW TO MEMORIZE FORMULAS

There are some handy but straightforward techniques/tools, which will help the students to improve their skills in learning math formulas. These are the general aspects of formula learning and applicable to all branches of Mathematics.

## *1. Practice and Revision*

Practice and Revision are the two most valuable tools to learn any formula. Students should work on several practice problems using one particular formula. The application of formulas always helps with memorization. Students cannot learn the formula by having one look only during the exam. They need to revise the formulas often during their free time preferably before they go to bed. Sticky notes are of great help in proper revision. Students can also make use of digital methods to learn the formula. Many videos available on various websites provide attractive graphics and images of the necessary formulas.

Math teachers are very unpopular amongst students as they give much homework. However, very few students know that they do it because they know that repetition is a critical aspect of learning. Practicing a new skill strengthened the connections between neurons in the brain. However, if students do not practice, then the weak bonds are likely to be broken. So, if the students try to learn formulas without doing the practice first, then they are just making it harder for themselves.

## 2. *Familiarize with the topic in advance*

Students tend to go through any lesson in mathematics only after a teacher teaches it in the school. However, it is a good idea to read over the upcoming lessons in the textbook before the teacher covers it in class. This advance reading for a few minutes does not mean that students need to memorize the formula they come across. It says that they are now a little more familiar with the topic to be discussed. This technique will also give them an overview of the diagrams, graphs, and vocabulary used in the new lesson.

Though, this step may take 15 minutes or so before each class but will make an enormous difference to students' understanding of the math they are studying.

Students who read a topic in advance, before the teacher discussed it in the school, used to remain calm in the class as compared to some of their batch mates who may be stressed out and confused about the new topic.

## 3. *Don't try to mug up lists of formulas*

Students are advised not to sit at their study table and attempt to learn the list of formula without fully understanding the meaning behind them. This practice will not only increase the amount of time it takes to learn but will also create lots of confusion. Moreover, if students can understand the connection between the meaning of each term used in the formula, their brain will be able to see links, and that will make them learn the formula quickly.

## *4. Know What and Why (Learning Derivations)*

Many students try to memorize the formulas without understanding the conditions attached to each formula, the meaning of the symbols used in the formula and the uses of a particular formula.

**Students should know, how a particular formula has arrived (learning derivation of the formula).** It helps in many cases, mainly when derivation of the formula involves few steps and it also works as an emergency tool if a student forgets a particular formula during examinations.

## 5. Make a Formula Book

Students should make a formula book for their reference on which they may write all necessary formulas. ***To make learning formulas easy, they may group the various formulas into lessons. They can further subdivide the formulas into subsections.*** This tool may work as an excellent filing system for the brain and help the brain to retrieve the formula at a faster pace.

For better revision, students can stick small notes on their bedroom walls. Sticky notes play a pivotal role during last minute revision for the exam.

## 6. Take Formula Tests Regularly

Students are advised to take Formula Test on a regular basis to remember various formulas. Students may ask their family members and friends to check the test and rate them accordingly, or they can do it themselves.

Frequent tests will not only boost the

confidence level of the students but also is an excellent revision technique. Practicing the problems using formulas that students need to know will make them understand the formula and would last in their memory for long.

Mental visualization is essential to check whether students remember the formulas or not. Students may close their eyes and try to recollect the formulas by saying it aloud. They may make a mental picture of the formula. Most formulas have a distinctive shape that students can remember like they would recognize a movie. This tool will help students to judge their remembrance power.

## 7. *Use the Right Memory Tools*

Most people can learn lists of unrelated numbers or words, as long as they use the proper memory tools. Students can apply such tools to learn the formulas as well.

One such powerful tool is "MNEMONICS." Some math formulas have mnemonics associated with it, which are in use for a long time. Usually, teachers use these to teach the students many formulas. Students can also use these mnemonic devices to learn formulas they are finding difficult.

For example, to learn the formulas for sine, cosine, and tangent in trigonometry, one can use the mnemonic like "SOH CAH TOA." Here, sine is opposite/ hypotenuse, cosine is adjacent /hypotenuse, and tangent is opposite /adjacent.

## 8. Never Ignore the Sleep

Parents should always caution students against underestimating the power of sleep when it comes to memorizing things. Our mind creates permanent memories each night during the deep sleep phase (called REM -Rapid Eye Movement Sleep). During this period, brain processes everything that one has learned during the day, and it rehearses all the new skills one has learned each day and started to consolidate them.

Students need to get enough sleep to be able to memorize things. They are, therefore, advised to avoid staying up late and cramming the night before an exam. The lack of sleep will cause stress, and they will not be able to perform well. It is better to plan their revision to have plenty of time before the examinations.

## 9. *Take care of health*

Students may find it a waste of time when they are busy studying; exercise is a great tool to get the oxygen flowing to their brain and can be beneficial to their learning. Also if students feel fit and healthy, they are less likely to be struck by stress-related illnesses which may hurt their studies.

Scientists have shown that the healthy students are less worried about sickness and thus not distracting them from learning. Students are therefore advised to spend some time exercising to getting the oxygen flowing into their brain.

## *10. Remove distractions*

Memorizing things (especially new lessons) requires full concentration. So students should try to avoid surfing the Internet, listening to music or texting their friends while studying. Many things can distract them from learning, so it is best to remove as many distractions as possible. Setting aside a period just for studying will help them to be more productive and will also make their learning more accessible.

# BY THE SAME AUTHOR

Readers would also like to read another book **(a bestseller at www.amazon.com)** titled **"AVOID SILLY MISTAKES IN MATHEMATICS."** The book is about various silly mistakes made by the students during school days (Age group 10-18 years). The book will not only help the students to overcome their habit of making silly mistakes but will also guide them in removing some of their misconceptions in mathematics.

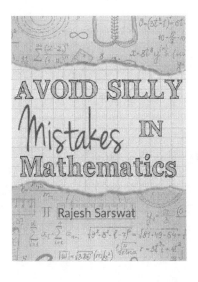

# 2
# TRICKS TO LEARN FORMULAS OF DERIVATIVES

Here is a list of some frequently used formulas of derivatives. In the list, alphabets a and n are the constants, e is the base of the natural logarithms, and u and v denote functions of x:

1. $\dfrac{d}{dx}(a) = 0$

2. $\dfrac{d}{dx}(x) = 1$

3. $\dfrac{d}{dx}(a.u) = a.\dfrac{du}{dx}$   where a = constant

4. $\dfrac{d}{dx}(u \pm v) = \dfrac{du}{dx} \pm \dfrac{dv}{dx}$

5. $\dfrac{d}{dx}(u.v) = u.\dfrac{dv}{dx} + v.\dfrac{du}{dx}$   (*Product Rule*)

6. $\dfrac{d}{dx}(u/v) = \dfrac{\left\{v\dfrac{du}{dx} - u\dfrac{dv}{dx}\right\}}{v^2}$ (*Quotient Rule*)

7. $\dfrac{d}{dx}\operatorname{Sin} x = \operatorname{Cos} x$

8. $\dfrac{d}{dx}\operatorname{Cos} x = -\operatorname{Sin} x$

9. $\dfrac{d}{dx}\tan x = \operatorname{Sec}^2 x$

10. $\dfrac{d}{dx}\operatorname{Cot} x = -\operatorname{Cosec}^2 x$

11. $\dfrac{d}{dx}\operatorname{Sec} x = \operatorname{Sec} x . \tan x$

12. $\dfrac{d}{dx}\operatorname{Cosec} x = -\operatorname{Cosec} x . \operatorname{Cot} x$

13. $\dfrac{d}{dx}\sin^{-1} x = \dfrac{1}{\sqrt{1-x^2}}$

14. $\dfrac{d}{dx}\cos^{-1} x = \dfrac{-1}{\sqrt{1-x^2}}$

15. $\dfrac{d}{dx}\tan^{-1} x = \dfrac{1}{1+x^2}$

16. $\dfrac{d}{dx}\cot^{-1} x = \dfrac{-1}{1+x^2}$

## How to Memorize Formulas in Mathematics

17. $\dfrac{d}{dx}\sec^{-1} x = \dfrac{1}{x\sqrt{x^2 - 1}}$

18. $\dfrac{d}{dx}\operatorname{cosec}^{-1} x = \dfrac{-1}{x\sqrt{x^2 - 1}}$

19. $\dfrac{d}{dx} x^n = nx^{n-1}$

20. $\dfrac{d}{dx} e^x = e^x$

21. $\dfrac{d}{dx} a^x = a^x \cdot \log_e a$

22. $\dfrac{d}{dx} \log_e |x| = 1/x$

23. If $x = f(y)$, then

$$\dfrac{dy}{dx} = \dfrac{1}{dx/dy}$$

24. If $y = f(u)$ and $u = g(x)$ then

$$\dfrac{dy}{dx} = \dfrac{dy}{du} \cdot \dfrac{du}{dx} \text{ (Chain Rule)}$$

25. If $y = f(t)$ and $x = g(t)$ then

$$\dfrac{dy}{dx} = \dfrac{f'(t)}{g'(t)} \text{ (Parametric Form)}$$

## Trick-1 Grouping of Formulas

As explained in chapter-1, divide all the twenty-five formulas into six groups as given hereunder. This simple strategy will not only help students to identify the pattern of the formulas in a particular group closely but also reduce psychological pressure from their mind. This step will also enable them to make a sort of filing system for their brain from where they can retrieve a particular formula as and when required.

## Group-1 Basic Properties of Derivative

1. $\dfrac{d}{dx}(a) = 0$

2. $\dfrac{d}{dx}(x) = 1$

3. $\dfrac{d}{dx}(a.u) = a.\dfrac{du}{dx}$

4. $\dfrac{d}{dx}(u \pm v) = \dfrac{du}{dx} \pm \dfrac{dv}{dx}$

How to Memorize Formulas in Mathematics

5. $\dfrac{d}{dx}(u.v) = u.\dfrac{dv}{dx} + v.\dfrac{du}{dx}$ (*Product Rule*)

6. $\dfrac{d}{dx}(u/v) = \dfrac{\left\{v\dfrac{du}{dx} - u\dfrac{dv}{dx}\right\}}{v^2}$ (*Quotient rule*)

7. If $x = f(y)$, then

$\dfrac{dy}{dx} = \dfrac{1}{dx/dy}$

8. If $y = f(u)$ and $u = g(x)$ then

$\dfrac{dy}{dx} = \dfrac{dy}{du}.\dfrac{du}{dx}$ (*Chain Rule*)

9. If $y = f(t)$ and $x = g(t)$ then

$\dfrac{dy}{dx} = \dfrac{f'(t)}{g'(t)}$ (*Parametric Form*)

## *Group-2 Trigonometric Functions*

10. $\dfrac{d}{dx} \sin x = \cos x$

11. $\dfrac{d}{dx} \cos x = -\sin x$

12. $\dfrac{d}{dx} \tan x = \sec^2 x$

13. $\dfrac{d}{dx} \cot x = -\csc^2 x$

14. $\dfrac{d}{dx} \sec x = \sec x \cdot \tan x$

15. $\dfrac{d}{dx} \operatorname{Cosec} x = - \operatorname{Cosec} x \cdot \operatorname{Cot} x$

## Group-3 Inverse Trigonometric Functions

16. $\dfrac{d}{dx} \sin^{-1} x = \dfrac{1}{\sqrt{1 - x^2}}$

17. $\dfrac{d}{dx} \cos^{-1} x = \dfrac{-1}{\sqrt{1 - x^2}}$

18. $\dfrac{d}{dx} \tan^{-1} x = \dfrac{1}{1 + x^2}$

19. $\dfrac{d}{dx} \cot^{-1} x = \dfrac{-1}{1 + x^2}$

20. $\dfrac{d}{dx} \sec^{-1} x = \dfrac{1}{x\sqrt{x^2 - 1}}$

21. $\dfrac{d}{dx} \operatorname{cosec}^{-1} x = \dfrac{-1}{x\sqrt{x^2 - 1}}$

## Group-4 Other Functions

22. $\dfrac{d}{dx} x^n = n x^{n-1}$

23. $\dfrac{d}{dx} e^x = e^x$

24. $\dfrac{d}{dx} a^x = a^x . \log_e a$

25. $\dfrac{d}{dx} \log_e |x| = 1/x$

## *Trick-2 For Negative Derivatives*

Students often commit mistakes while dealing with functions having negative derivatives. Most of the students are confused with derivatives of some functions that have a negative derivative.

Students are advised to go through the 16 formulas of Group 2, Group 3 and Group 4 as stated on previous pages and observe the functions which have negative derivatives. They will found these functions as:

1. $\dfrac{d}{dx} \text{Cos } x = -\text{Sin } x$

2. $\dfrac{d}{dx} \text{Cot } x = -\text{Cosec}^2 x$

3. $\dfrac{d}{dx} \text{Cosec } x = -\text{Cosec } x . \text{Cot } x$

4. $\dfrac{d}{dx}\cos^{-1} x = \dfrac{-1}{\sqrt{1-x^2}}$

5. $\dfrac{d}{dx}\cot^{-1} x = \dfrac{-1}{1+x^2}$

6. $\dfrac{d}{dx} co\sec^{-1} x = \dfrac{-1}{x\sqrt{x^2-1}}$

A quick observation conveys that **all the functions of Group 2 and Group 3 starting with the letter 'C' have negative derivatives.**

## *Trick-3 Formulas of Group 2*

Students can observe at a glance that Sin x and Cos x are derivatives of each other, and as Cos x is starting with the letter **'C' (Trick 2)**, it will have a negative derivative. So, we have:

$\dfrac{d}{dx}\text{Sin x} = \text{Cos x}$

$\dfrac{d}{dx}\text{Cos x} = -\text{Sin x}$

For learning the derivative of the remaining four trigonometric functions,

How to Memorize Formulas in Mathematics

students can learn the following table:

| Tan x | Sec x | Sec x |
|---|---|---|
| Cot x | -Cosec x | Cosec x |

Now let us see, how does this table work?

| Tan x | Sec x | Sec x |
|---|---|---|
| Cot x | -Cosec x | Cosec x |

To find the derivative of tan x, students should multiply the remaining two terms other than tan x (first term) of the first row to get derivative as $\sec^2 x$.

So, $\frac{d}{dx} \tan x = \sec^2 x$

| Tan x | Sec x | Sec x |
|---|---|---|
| Cot x | -Cosec x | Cosec x |

To find the derivative of sec x, students should multiply the remaining two terms other than other than sec x ( last term) to get derivative as sec x . tanx

So, $\frac{d}{dx} \sec x = \sec x . \tan x$

Similarly, students can find the derivative of Cosec x and Cot x by using the second row as follows:

| Tan x | Sec x | Sec x |
|-------|-------|-------|
| Cot x | -Cosec x | **Cosec x** |

$So, \dfrac{d}{dx} \text{Cosec } x = - \text{Cosec } x . \text{Cot } x$

| Tan x | Sec x | Sec x |
|-------|-------|-------|
| **Cot x** | -Cosec x | Cosec x |

$So, \dfrac{d}{dx} \text{Cot } x = -\text{Cosec}^2 x$

Students should learn the above table because the same will again be used to memorize more results on integration in Chapter-3.

## *Trick-4 Formulas of Group 3*

Students can learn these formulas by dividing these results into two sub-groups of three formulas each as tabulated on next page:

How to Memorize Formulas in Mathematics

| Group-I | Group-II |
|---|---|
| $\dfrac{d}{dx}\sin^{-1} x = \dfrac{1}{\sqrt{1-x^2}}$ | $\dfrac{d}{dx}\cos^{-1} x = \dfrac{-1}{\sqrt{1-x^2}}$ |
| $\dfrac{d}{dx}\tan^{-1} x = \dfrac{1}{1+x^2}$ | $\dfrac{d}{dx}\cot^{-1} x = \dfrac{-1}{1+x^2}$ |
| $\dfrac{d}{dx}\sec^{-1} x = \dfrac{1}{x\sqrt{x^2-1}}$ | $\dfrac{d}{dx}\cosec^{-1} x = \dfrac{-1}{x\sqrt{x^2-1}}$ |

Students may observe the following things to memorize these formulas:

1. Formulae for $\sin^{-1}x$ and $\cos^{-1}x$ are the same except for the negative sign in the formula of $\cos^{-1}x$ **(trick -2)**.

2. Formulae for $\tan^{-1}x$ and $\cot^{-1}x$ same except for the negative sign in the formula of $\cot^{-1}x$ **(trick -2)**.

3. Formulae for $\sec^{-1}x$ and $\cosec^{-1}x$ are

the same except for the negative sign in the formula of $\mathrm{cosec}^{-1}x$ **(trick -2).**

4. Students can learn the formulas of $\sin^{-1}x$ and $\sec^{-1}x$ **(starting with letter 'S' )** by linking 'S' for **'Subtraction'** and **'S'** for **'Square Root'** with the formulas as both the formulas have a subtraction and square root.

5. For learning formula of $\tan^{-1}x$ students may link the first letter of **t**an, which looks like a **+** sign, with a + sign in its formula.

6. In a nutshell, formulas of **S**in$^{-1}$x and **S**ec$^{-1}$x will have **S**ubtraction inside **S**quare Root as these functions are starting with ' **S.**' Formula of **t**an$^{-1}$x have a + sign in it as '+' sign resembles with the letter ' **t.**' Also, it does not have a square root in it as it does not start with the letter **'S.'**

# 3
# TRICKS TO LEARN FORMULAS OF INTEGRATION

A list of some basic formulas of Integration is as under:

1. $\int x^n \, dx = \dfrac{x^{n+1}}{n+1} + C$

2. $\int e^x \, dx = e^x + C$

3. $\int a^x \, dx = \dfrac{a^x}{\log_e a} + C$

4. $\int 1/x \, dx = \log_e |x| + C$

5. $\int \sin x \, dx = -\cos x + C$

6. $\int \cos x \, dx = \sin x + C$

7. $\int \sec^2 x \, dx = \tan x + C$

8. $\int \text{Cosec}^2 x \, dx = -\text{Cot } x + C$

9. $\int \text{Sec } x \cdot \tan x \, dx = \text{Sec } x + C$

10. $\int \text{Cosec } x \cdot \text{Cot } x \, dx = -\text{Cosec } x + C$

11. $\int \text{Tan } x \, dx = -\log_e |\text{Cos } x| + C$

12. $\int \text{Cot } x \, dx = \log_e |\text{Sin } x| + C$

13. $\int \text{Sec } x \, dx = \log_e |\text{Sec } x + \tan x| + C$

14. $\int \text{Cosec } x \, dx = \log_e |\text{Cosec } x - \text{Cot } x| + C$

15. $\int \dfrac{1}{\sqrt{1-x^2}} \, dx = \sin^{-1} x + C$

16. $\int \dfrac{-1}{\sqrt{1-x^2}} \cos^{-1} x \, dx = \cos^{-1} x + C$

17. $\int \dfrac{1}{1+x^2} \, dx = \tan^{-1} x + C$

18. $\int \dfrac{-1}{1+x^2} \, dx = \cot^{-1} x + C$

19. $\int \dfrac{1}{x\sqrt{x^2-1}} \, dx = \sec^{-1} x + C$

# How to Memorize Formulas in Mathematics

20. $\int \dfrac{-1}{x\sqrt{x^2-1}} \, dx = \cosec^{-1} x + C$

21. $\int \sqrt{a^2 - x^2} \, dx$

$= \dfrac{x}{2}\sqrt{a^2 - x^2} + \dfrac{a^2}{2} \sin^{-1}\dfrac{x}{a} + C$

22. $\int \sqrt{x^2 - a^2} \, dx$

$= \dfrac{x}{2}\sqrt{x^2 - a^2} - \dfrac{a^2}{2} \log |x + \sqrt{x^2 - a^2}| + C$

23. $\int \sqrt{x^2 + a^2} \, dx$

$= \dfrac{x}{2}\sqrt{x^2 + a^2} + \dfrac{a^2}{2} \log |x + \sqrt{x^2 + a^2}| + C$

24. $\int \dfrac{1}{\sqrt{x^2 + a^2}} \, dx = \log |x + \sqrt{x^2 + a^2}| + C$

25. $\int \dfrac{1}{\sqrt{x^2 - a^2}} \, dx = \log |x + \sqrt{x^2 + a^2}| + C$

26. $\int \dfrac{1}{\sqrt{a^2 - x^2}} \, dx = \sin^{-1}\dfrac{x}{a} + C$

27. $\int \dfrac{1}{x^2 + a^2} \, dx = \dfrac{1}{a} \tan^{-1}\dfrac{x}{a} + C$

28. $\int \dfrac{1}{x^2 - a^2} \, dx = \dfrac{1}{2a} \log_e \dfrac{x - a}{x + a} + C$

29. $\int \dfrac{1}{a^2 - x^2} \, dx = \dfrac{1}{2a} \log_e \dfrac{a+x}{a-x} + C$

## *Trick-1 Grouping of Formulas:*

As explained in chapter-1, divide all the twenty- nine formulas into six groups as given hereunder. This simple act will help students to identify the pattern of all these formulas closely.

## *Group-1*

1. $\int x^n \, dx = \dfrac{x^{n+1}}{n+1} + C$

2. $\int e^x \, dx = e^x + C$

3. $\int a^x \, dx = \dfrac{a^x}{\log_e a} + C$

4. $\int 1/x \, dx = \log_e |x| + C$

## *Group-2*

5. $\int \sin x \, dx = -\cos x + C$

6. $\int \cos x \, dx = \sin x + C$

7. $\int \sec^2 x \, dx = \tan x + C$

8. $\int \csc^2 x \, dx = -\cot x + C$

9. $\int \sec x \cdot \tan x \, dx = \sec x + C$

10. $\int \csc x \cdot \cot x \, dx = -\csc x + C$

## Group-3
### 3 (A)

11. $\int \tan x \, dx = -\log_e |\cos x| + C$

12. $\int \cot x \, dx = \log_e |\sin x| + C$

### 3 (B)

13. $\int \sec x \, dx = \log_e |\sec x + \tan x| + C$

14. $\int \csc x \, dx = \log_e |\csc x - \cot x| + C$

## Group-4

15. $\int \dfrac{1}{\sqrt{1-x^2}} \, dx = \sin^{-1} x + C$

16. $\int \dfrac{-1}{\sqrt{1-x^2}} \cos^{-1} x \, dx = \cos^{-1} x + C$

17. $\int \dfrac{1}{1+x^2}.dx = \tan^{-1} x + C$

18. $\int \dfrac{-1}{1+x^2}.dx = \cot^{-1} x + C$

19. $\int \dfrac{1}{x\sqrt{x^2-1}}.dx = \sec^{-1} x + C$

20. $\int \dfrac{-1}{x\sqrt{x^2-1}}.dx = \operatorname{cosec}^{-1} x + C$

## *Group-5*

21. $\int \sqrt{a^2 - x^2}.dx$

$= \dfrac{x}{2}\sqrt{a^2 - x^2} + \dfrac{a^2}{2}\sin^{-1}\dfrac{x}{a} + C$

22. $\int \sqrt{x^2 - a^2}.dx$

$= \dfrac{x}{2}\sqrt{x^2 - a^2} - \dfrac{a^2}{2}\log |x + \sqrt{x^2 - a^2}| + C$

23. $\int \sqrt{x^2 + a^2}.dx$

$= \dfrac{x}{2}\sqrt{x^2 + a^2} + \dfrac{a^2}{2}\log |x + \sqrt{x^2 + a^2}| + C$

## Group-6
### 6 (A)

24. $\int \dfrac{1}{\sqrt{x^2 + a^2}} \, dx = \log |x + \sqrt{x^2 + a^2}| + C$

25. $\int \dfrac{1}{\sqrt{x^2 - a^2}} \, dx = \log |x + \sqrt{x^2 - a^2}| + C$

### 6 (B)

26. $\int \dfrac{1}{\sqrt{a^2 - x^2}} \, dx = \sin^{-1}\dfrac{x}{a} + C$

27. $\int \dfrac{1}{x^2 + a^2} \, dx = \dfrac{1}{a}\tan^{-1}\dfrac{x}{a} + C$

### 6 (C)

28. $\int \dfrac{1}{x^2 - a^2} \, dx = \dfrac{1}{2a}\log_e \dfrac{x - a}{x + a} + C$

29. $\int \dfrac{1}{a^2 - x^2} \, dx = \dfrac{1}{2a}\log_e \dfrac{a + x}{a - x} + C$

## Trick 2 Formulas of Group-1

Integration is the reverse process as of differentiation. By using this fact, students can easily memorize two out of four formulas in this group:

1. $\int e^x . dx = e^x + C$

as $\dfrac{d}{dx} e^x = e^x$

2. $\int \dfrac{1}{x} . dx = \log_e |x| + C$

as $\dfrac{d}{dx} \log_e |x| = 1/x$

(While getting the formulas of Integration, the right side of the derivative formula becomes the left side, and the left side of the derivative formula becomes the right side).

3. Thus, $\dfrac{d}{dx} a^x = a^x . \log_e a$

**Leads to the result:**

$\int a^x . \log_e a . dx = a^x$

or $\int a^x . dx = \dfrac{a^x}{\log_e a} + C$

4. $\dfrac{d}{dx} x^n = nx^{n-1}$

**Leads to the result :**

$$\frac{d}{dx} x^{n+1} = (n+1) \cdot x^n$$

**Or**

$$\int (n+1) \cdot x^n \, dx = x^{n+1}$$

which further gives,

$$\int x^n \, dx = \frac{x^{n+1}}{(n+1)} + C$$

## Trick 3 Formulas of Group-2

Again, students can learn all the formulas of this group by using the formulas of derivative in reverse order as follows:

5. As, $\dfrac{d}{dx} Cos\, x = -Sin\, x$

or $\dfrac{d}{dx}(-Cos\, x) = Sin\, x$

$$\int Sin\, x \, dx = -Cos\, x + C$$

6. As, $\dfrac{d}{dx} Sin\, x = Cos\, x$

$$\int Cos\, x \, dx = Sin\, x + C$$

7. As $\frac{d}{dx} \tan x = Sec^2 x$

$\int Sec^2 x \, . \, dx = \tan x + C$

8. As $\frac{d}{dx} Cot \, x = -Cosec^2 x$

or $\frac{d}{dx}(-Cot \, x) = Cosec^2 x$

$\int Cosec^2 x \, . \, dx = -Cot \, x + C$

9. As $\frac{d}{dx} Sec \, x = Sec \, x \, . \, \tan x$

$\int Sec \, x \, . \, \tan x \, . \, dx = Sec \, x + C$

10. As $\frac{d}{dx} Cosec \, x = - \, Cosec \, x \, . \, Cot \, x$

or $\frac{d}{dx}(-Cosec \, x) = Cosec \, x \, . \, Cot \, x$

$\int Cosec \, x \, . \, Cot \, x \, . \, dx = -Cosec \, x + C$

Formula No 7, 8, 9, 10 can also be memorized by using the following table as used in Chapter 2.

## How to Memorize Formulas in Mathematics

| Tan x | Sec x   | Sec x   |
|-------|---------|---------|
| Cot x | -Cosec x| Cosec x |

| Tan x | Sec x   | Sec x   |
|-------|---------|---------|
| Cot x | -Cosec x| Cosec x |

$\int Sec^2 x \, dx = \tan x + C$ ( from Row 1)

| Tan x | Sec x   | Sec x   |
|-------|---------|---------|
| Cot x | -Cosec x| Cosec x |

$\int Sec\, x . \tan x \, dx = Sec\, x + C$ (From Row 1)

| Tan x | Sec x   | Sec x   |
|-------|---------|---------|
| Cot x | -Cosec x| Cosec x |

$\int -Cosec^2 x \, dx = Cot\, x + C$ (From Row $-$ 2)

| Tan x | Sec x   | Sec x   |
|-------|---------|---------|
| Cot x | -Cosec x| Cosec x |

$or \int Cosec^2 x \, dx = -Cot\, x + C$

$\int -Cosec\, x . Cot\, x \, dx = Cosec\, x + C$ *or*

$\int Cosec\, x . Cot\, x \, dx = -Cosec\, x + C$ **(From Row-2)**

# *HOW TO MEMORIZE FORMULAS OF GROUP-3*

Students should remember that the formulas of Integration of tan x, cot x, sec x, and cosec x contain logarithm function in the answer.

## *GROUP-3 (A)*

The best ways to learn these two formulas is the derivation of these two, which is very simple to learn.

11. $\int \tan x \, dx = \int \dfrac{\sin x}{\cos x} \, dx = - \int \dfrac{1}{t} \, dt$

$\qquad = -\log_e |t| + C = -\log_e |\cos x| + C$

**(Substitute cos x = t or – sin x. dx = dt)**

12. $\int \cot \, dx = \int \dfrac{\cos x}{\sin x} \, dx = \int \dfrac{1}{t} \, dt$

$\qquad = \log_e |t| + C = \log_e |\sin x| + C$

**(Substitute sin x = t or cos x. dx = dt)**

## *GROUP-3 (B)*

For learning the formulas of this group,

again refer back to the following table:

| Tan x | Sec x | Sec x |
|---|---|---|
| Cot x | -Cosec x | Cosec x |

13. $\int \sec x \, dx = \log_e |\sec x + \tan x| + C$

14. $\int \text{Cosec } x \, dx = \log_e |\cot x - \text{cosec } x| + C$

**Or**

$\int \text{Cosec } x \, dx = \log_e |\text{cosec } x - \cot x| + C$

**(as $|a - b| = |b - a|$ )**

## *HOW TO MEMORIZE FORMULAS OF GROUP-4*

All the formulas in this group can easily be memorized with the help of the formulas of derivatives of Inverse Trigonometric Functions as follows:

15. $\int \dfrac{1}{\sqrt{1-x^2}} \, dx = \sin^{-1} x + C$

( As $\dfrac{d}{dx} \sin^{-1} x = \dfrac{1}{\sqrt{1-x^2}}$ )

16. $\int \dfrac{-1}{\sqrt{1-x^2}} \cos^{-1} x \, . \, dx = \cos^{-1} x + C$

$(As \; \dfrac{d}{dx} \cos^{-1} x = \dfrac{-1}{\sqrt{1-x^2}})$

17. $\int \dfrac{1}{1+x^2} . \, dx = \tan^{-1} x + C$

$(As \; \dfrac{d}{dx} \tan^{-1} x = \dfrac{1}{1+x^2})$

18. $\int \dfrac{-1}{1+x^2} . \, dx = \cot^{-1} x + C$

$(As \; \dfrac{d}{dx} \cot^{-1} x = \dfrac{-1}{1+x^2})$

19. $\int \dfrac{1}{x\sqrt{x^2-1}} . \, dx = \sec^{-1} x + C$

$(As \; \dfrac{d}{dx} \sec^{-1} x = \dfrac{1}{x\sqrt{x^2-1}})$

20. $\int \dfrac{-1}{x\sqrt{x^2-1}} . \, dx = \cosec^{-1} x + C$

$(As \; \dfrac{d}{dx} co\sec^{-1} x = \dfrac{-1}{x\sqrt{x^2-1}})$

## *HOW TO MEMORIZE FORMULAS OF GROUP-5*

The integration of these functions has two

parts. Observe the first part of the integration by using the following table:

| Formula | First Part of the Answer | Observation |
|---|---|---|
| 21. $\int \sqrt{a^2 - x^2} \, dx$ | $\dfrac{x}{2}\sqrt{a^2 - x^2}$ | The first part of the answer in all the cases is x/2 multiplied by the function itself. |
| 22. $\int \sqrt{x^2 - a^2} \, dx$ | $\dfrac{x}{2}\sqrt{x^2 - a^2}$ | |
| 23. $\int \sqrt{x^2 + a^2} \, dx$ | $\dfrac{x}{2}\sqrt{x^2 + a^2}$ | |

Observe the second part of the integration by using the following table:

| Formula | Second Part of the Answer |
|---|---|
| 21. $\int \sqrt{a^2 - x^2} \, dx$ | $\dfrac{a^2}{2} \sin^{-1}\dfrac{x}{a}$ |
| 22. $\int \sqrt{x^2 - a^2} \, dx$ | $-\dfrac{a^2}{2} \log |x + \sqrt{x^2 - a^2}|$ |
| 23. $\int \sqrt{x^2 + a^2} \, dx$ | $\dfrac{a^2}{2} \log |x + \sqrt{x^2 + a^2}|$ |

## Observations in the second part of the answer:

1. The second part of each formula contains $a^2/2$, and the sign of $a^2/2$ in the answer is same as the sign of $a^2$ in the function.

2. Answers of Formula 22 and 23 contain a term in Logarithm, i.e. log | x + Function itself |.

3. Formula No. 21 and 26 have $\sqrt{a^2 - x^2}$ in them, link it with the term $\sin^{-1}\frac{x}{a}$.

## HOW TO MEMORIZE FORMULAS OF GROUP-6

**Students should observe the pattern of these formulas and may learn these results after dividing them into three subgroups as follows:**

## 6 (A)

The answer to both these formulas is log | x + Function in the denominator |.

24. $\int \dfrac{1}{\sqrt{x^2+a^2}} \, dx = \log |x + \sqrt{x^2+a^2}| + C$

(Second Part of the formula number 23 excluding $a^2/2$)

25. $\int \dfrac{1}{\sqrt{x^2-a^2}} \, dx = \log |x + \sqrt{x^2-a^2}| + C$

(Second Part of the formula number 22 excluding $-a^2/2$)

## 6 (B)

26. $\int \dfrac{1}{\sqrt{a^2-x^2}} \, dx = \sin^{-1}\dfrac{x}{a} + C$

(Second Part of the formula number 21 excluding $a^2/2$)

27. $\int \dfrac{1}{x^2+a^2} \, dx = \dfrac{1}{a}\tan^{-1}\dfrac{x}{a} + C$

(Recall formula number 17 and divide it by a twice, once before $\tan^{-1}$ and then after $\tan^{-1}$).

## 6 (C)

The answer to both these formulas is of the form

$$\frac{1}{2a}\log_e \left|\frac{Factor-1\ of\ denominator}{Factor-2\ of\ denominator}\right|$$

28. $\int \frac{1}{x^2 - a^2} \cdot dx = \frac{1}{2a}\log_e \left|\frac{x-a}{x+a}\right| + C$

(Factors of $x^2 - a^2$ are $(x - a)$ and $(x + a)$ but factor with negative a is in the numerator as the sign of $a^2$ is negative)

29. $\int \frac{1}{a^2 - x^2} \cdot dx = \frac{1}{2a}\log_e \left|\frac{a+x}{a-x}\right| + C$

(Factors of $a^2 - x^2$ are $(a - x)$ and $(a + x)$ but factor with positive a is in the numerator as the sign of $a^2$ is positive).

# 4
# SILLY MISTAKES IN CALCULUS

## i. Ignoring Notations for Limits

Students often forget to write the notations for limits while solving these questions after few steps as evident from the following example:

$$\lim_{x \to 3} \frac{x^2 - 9}{x - 3} = \frac{(x - 3)(x + 3)}{x - 3} = x + 3 = 6$$

The correct way of writing this solution should be as under:

$$\lim_{x \to 3} \frac{x^2 - 9}{x - 3} = \lim_{x \to 3} \frac{(x - 3)(x + 3)}{x - 3} = \lim_{x \to 3} (x + 3) = 6$$

**Hence, students should keep writing the notation for limits up to the step in which they substitute the value of limit to the given function.**

## ii. Applying the limits on the part of a function

Students often use the value of limit on the part of function without simplifying the function altogether. It will be evident from the following example:

$$\lim_{x \to 0} \frac{ax + x \cos x}{b \sin x} = \frac{1}{b} \lim_{x \to 0} \frac{x(a + \cos x)}{\sin x}$$

$$= \frac{1}{b} \lim_{x \to 0} \frac{x(a + \cos 0)}{\sin x} = \frac{1}{b} \lim_{x \to 0} \frac{x(a + 1)}{\sin x}$$

$$= \frac{1}{b} \lim_{x \to 0} \frac{x}{\sin x} \cdot (a + 1) = \frac{a+1}{b} \lim_{x \to 0} \frac{x}{\sin x} = \frac{a+1}{b}$$

$$\left\{ \because \lim_{x \to 0} \frac{x}{\sin x} = 1 \right\}$$

In the above example, the student has applied the limit to some part of the Numerator without simplifying the entire function. The correct solution should be as under:

$$\lim_{x \to 0} \frac{ax + x\cos x}{b\ sinx} = \frac{1}{b} \lim_{x \to 0} \frac{x(a + \cos x)}{sinx}$$

$$= \frac{1}{b} \lim_{x \to 0} \frac{x(a + \cos x)}{sinx} = \frac{1}{b} \lim_{x \to 0} \frac{x}{sinx} \cdot (a + cosx)$$

$$= \frac{1}{b} \cdot 1 \cdot (a + \cos 0) = \frac{a + 1}{b}$$

### iii. *Improper use of L' Hospital's Rule*

**L' Hospital's Rule:** As per this rule, for the limits of the following types:

$$\lim_{x \to a} \frac{f(x)}{g(x)} = \frac{0}{0} \ \ or \ \ \lim_{x \to a} \frac{f(x)}{g(x)} = \frac{\pm \infty}{\pm \infty}$$

Where $a$ can be any real number, infinity or negative infinity, we have:

$$\lim_{x \to a} \frac{f(x)}{g(x)} = \lim_{x \to a} \frac{f'(x)}{g'(x)}$$

Students tend to use this formula as:

$$\lim_{x \to a} [f(x) \cdot g(x)] = \lim_{x \to a} [f'(x) \cdot g'(x)]$$

and

$$\lim_{x \to a}[f(x) \pm g(x)] = \lim_{x \to a}[f'(x) \pm g'(x)]$$

The above interpretations are wrong as L' Hospital's Rule is applicable only on the quotient of two functions and that too under certain conditions as specified on the previous page and not on the product, sum or difference of two functions.

### iv. *Improper use of the formula of derivative of $x^n$,*

$$\frac{d}{dx}x^n = nx^{n-1}$$

Students make a common mistake in differential calculus by misusing the above result. The students tend to forget that the above result is used only when the base number x is a variable and the exponent n is a constant (scalar) and not the vice versa and in any other similar looking functions.

## How to Memorize Formulas in Mathematics

**For example:**

$\frac{d}{dx}x^5 = 5x^{5-1} = 5x^4$ is alright, but,

$\frac{d}{dx}x^x = xx^{x-1} = x^x$ is wrong as exponent x is a variable. The correct way of solving problems like this is as follows:

Let $y = x^x$,

$\log_e y = \log_e x^x$

$\log_e y = x \cdot \log_e x$

$\frac{d}{dx}\log_e y = \frac{d}{dx}x \cdot \log_e x$

$\frac{1}{y}\frac{dy}{dx} = x\frac{d}{dx}\log_e x + \log_e x \frac{d}{dx}x$

$\frac{1}{y}\frac{dy}{dx} = x\frac{d}{dx}\log_e x + \log_e x \cdot 1$

$\frac{1}{y}\frac{dy}{dx} = 1 + \log_e x$

$\frac{d}{dx}x^x = x^x \cdot (1 + \log_e x)$

Similarly,

$\frac{d}{dx}e^x = x \cdot e^{x-1}$ and

$\frac{d}{dx}a^x = x.a^{x-1}$, are wrong as in these examples, the exponent x is a variable, and the base numbers e and a are constants. The correct formulae for derivative of these functions are:

$$\frac{d}{dx}e^x = e^x$$

and

$$\frac{d}{dx}a^x = a^x.\log_e a$$

Therefore,

$$\frac{d}{dx}5^x = 5^x.\log_e 5$$

**Important Note:** When the base is a variable, and the exponent is a constant, students should use the following formula:

$$\frac{d}{dx}x^n = n.x^{n-1}$$

However, when the base is a constant and the exponent is a variable, students should use the following formula:

How to Memorize Formulas in Mathematics

$$\frac{d}{dx}a^x = x \cdot a^{x-1}$$

There is a separate formula for

$\frac{d}{dx}e^x = e^x$, where e is a constant but it is a special constant and so is separate from a as used in $\frac{d}{dx}a^x$.

Students will learn the difference between $a^x$ and $e^x$ in future topics.

### v. *Finding derivative of [f(x).g(x)] incorrectly*

The property for derivative of Sum or difference of two functions is as under:

$$\frac{d}{dx}[f(x) \pm g(x)] = \frac{d}{dx}f(x) \pm \frac{d}{dx}g(x)$$

However, students often try to apply the same analogy to find the derivative of the product of two functions as:

$\frac{d}{dx}[f(x) \cdot g(x)] = \frac{d}{dx}f(x)\frac{d}{dx}g(x)$, which is wrong.

The correct formula for finding derivative of the product of two functions is:

$$\frac{d}{dx}[f(x).g(x)] = g(x)\frac{d}{dx}f(x) + f(x)\frac{d}{dx}g(x)$$

## vi. *Finding derivative of [f(x)/g(x)] incorrectly*

The property for derivative of Sum or difference of two functions is as under:

$$\frac{d}{dx}[f(x) \pm g(x)] = \frac{d}{dx}f(x) \pm \frac{d}{dx}g(x)$$

However, students often try to apply the same analogy to find derivative of the quotient of two functions as:

$$\frac{d}{dx}[f(x)/g(x)] = \frac{d}{dx}f(x)/\frac{d}{dx}g(x) \quad, \text{ which is wrong.}$$

**The correct formula for finding derivative of the product of two functions is:**

$$\frac{d}{dx}\left[\frac{f(x)}{g(x)}\right] = \left\{g(x)\frac{d}{dx}f(x) - f(x)\frac{d}{dx}g(x)\right\}/\{g(x)\}^2$$

## How to Memorize Formulas in Mathematics

### *vii. Ignoring constants in Integration*

Integration is the inverse process that of finding a derivative.

Thus if,

$\frac{d}{dx}(\text{Sin } x) = \cos x$,

$\int \cos x \, dx = \sin x$.

That means, if the derivative of sin x with respect to x is cos x, then, integration of cos x with respect to x will be sin x.

However,

$\frac{d}{dx}(\text{Sin } x)$

$= \frac{d}{dx}(\text{Sin } x + 2)$

$= \frac{d}{dx}(\text{Sin } x - 10)$ …. and so on $= \cos x$

That yields, $\int \cos x \, dx = \sin x$ or sin x + 2 or sin x −10, which shows that integration of a function may yield infinite values. So, it will be more appropriate to write, ∫ cos x dx = sin x + c, where c is an arbitrary constant.

Due to ignorance or haste, students tend to drop the constant while finding the integration of some function which is not correct. Thus, ∫ cos x dx = sin x is wrong, and students should write it as ∫ cos x dx = sin x + c as there are infinite values of integration of a function.

## *viii. Using double constants in Integration*

In the last topic it has been re-iterated that while writing the answer for integration of a function, students need to add a constant to the solution to make the perfect sense. However, sometimes, students use two constants in the same question, which is not a good practice. For example: In the following question,

$$\int \frac{\sin x}{\sin(x-a)} \cdot dx$$

## How to Memorize Formulas in Mathematics

Let $x - a = t$ or $dx = dt$

$$\int \frac{\sin x}{\sin(x-a)} \cdot dx = \int \frac{\sin(t+a)}{\sin t} \cdot dt$$

$$= \int \frac{\sin t \cdot \cos a + \cos t \cdot \sin a}{\sin t} \cdot dt$$

$$= \int (\cos a + \cos t \cdot \sin a) \cdot dt$$

$$= t \cdot \cos a + \sin a \cdot \log_e |\sin t| + C_1$$
$$= (x - a) \cdot \cos a + \sin a \cdot \log_e |\sin(x - a)| + C_1$$
$$= x \cdot \cos a + \sin a \cdot \log_e |\sin(x - a)| - a\cos a + C_1$$

Leaving the answer at this stage will be against the convention as there are two constants in the answer, $-a \cos a$ and $C_1$.

The correct way should be to write the resultant of $-a \cos a$ and $C_1$ as $C$ and thus in the final solution, only one constant has been used. The answer should look like:

$$= x \cdot \cos a + \sin a \cdot \log_e |\sin(x - a)| + C$$

### ix. Improper use of the formula for $\int x^n \, dx$

Students often forget that there is a restriction on this integration formula, so the formula along with the restriction is under:

$\int x^n \cdot dx = \frac{x^{n+1}}{n+1}$,

provided n ≠ -1

Thus it is wrong to use the formula where n = -1, in the following case:

$\int \frac{1}{x} \cdot dx = \int x^{-1} \cdot dx = \frac{x^{-1+1}}{-1+1} + c = \frac{x^0}{0} + c$

The correct formula for finding the integration of 1/x is as follows:

$\int 1/x \cdot dx = \log_e |x| + c$

### x. Dropping the absolute value when integrating $\int 1/x \, dx$

The formula for finding Integration of 1/x is as under:

$\int 1/x \cdot dx = \log_e |x| + c$

How to Memorize Formulas in Mathematics

However, most of the students tend to forget the sign of absolute value (modulus) attached with x in the formula, which is very much required.

Though, it is true that we do not require the notation for absolute value in some cases such as:

$$\int \frac{2x}{x^2 + 7} \cdot dx = \log_e |x^2 + 7| + c$$
$$= \log_e (x^2 + 7) + c$$

In the previous example, the value of x² + 7 is positive. Therefore the use of absolute value notation has no meaning, and therefore its use is optional.

However, consider the following case,

$$\int \frac{2x}{x^2 - 7} \cdot dx = \log_e |x^2 - 7| + c$$

In this case, the value of x² - 7 may be positive or negative depending on the value of x, and so the use of absolute value notation is necessary.

Students are, therefore, advised to use the notation for absolute value in all cases wherever, the formula of $\int 1/x \, dx = \log_e |x| + c$ is used.

## xi. Improper use of formula ∫1/x dx

The formula for finding Integration of 1/x is as under:

$\int 1/x \, dx = \log_e |x| + c$

However, this formula has been used by students in the wrong manner as will be evident from the following examples:

$\int 1/x^2 \, dx = \log_e |x^2| + c$

$\int 1/\sin x \, dx = \log_e |\sin x| + c$

$\int 1/e^x \, dx = \log_e |e^x| + c$

Students may note that the above formula works only when the numerator is 1, and the denominator is x or a linear expression of x only as follows:

$\int 1/(x+2) \, dx = \log_e |x+2| + c$

$\int 1/(2x+3) \, dx = \dfrac{1}{2} \log_e |2x+3| + c$

## xii. Finding ∫[f(x).g(x)] dx incorrectly

The property for derivative of Sum or difference of two functions is as under:

$$\int [f(x) \pm g(x)] \, dx = \int f(x) \, dx \pm \int g(x) \, dx$$

However, students often try to apply the same analogy to find the integration of the product of two functions as:

$\int [f(x).g(x)] \, dx = \int f(x) \, dx . \int g(x) \, dx$ , which is wrong.

The method for finding the integration of the product of two functions is called Integration by parts, and its formula is given by:

$$\int [f(x).g(x)] \, dx$$

$$= f(x) \int g(x) dx - \int \left(\frac{d}{dx} f(x)\right) \left(\int g(x) dx\right) dx$$

For example, to get $\int x \cdot \sin x \, dx$

Let $I = \int x \cdot \sin x \, dx$. Taking $x$ as the first function and $\sin x$ as the second function and integrating by parts:

$$I = x \int \sin x \, dx - \int \left\{\left(\frac{d}{dx} x\right) \int \sin x \cdot dx\right\} \cdot dx$$

$$= x(-\cos x) - \int 1 \cdot (-\cos x) \, dx$$

$$= -x\cos x + \sin x + C$$

## *xiii. Finding $\int [f(x)/g(x)] \, dx$ incorrectly*

The property for derivative of Sum or difference of two functions is as under:

$$\int [f(x) \pm g(x)] \, dx = \int f(x) dx \pm \int g(x) dx$$

However, students often try to apply the same analogy to find integration of the quotient of two functions as:

$\int [f(x) / g(x)] \, dx = \int f(x) dx / \int g(x) dx$ ,which is wrong.

**There is no direct formula for finding the integration of the quotient of two functions and the methods of finding the integration in such cases vary from case to case basis.**

For evaluating

$\int \frac{2x}{1+x^2} . dx;$

Substitution can solve this problem s follows:

$Let, 1 + x^2 = t \text{ or}$

$2x . \frac{dx}{dt} = 1 \text{ or } 2x . dx = dt$

The given integration reduces to:

$\int \frac{2x}{1 + x^2} . dx = \int \frac{1}{t} . dt = \log_e |t| + c$

$= \log_e |1 + x^2| + c$

## xiv. Dealing with limits in Definite Integration

While attempting questions of definite

integration by way of substitution, sometimes limits are not changed to match the new variables, and that makes an error.

For example, for solving

$$\int_0^1 \frac{2x}{1+x^2} \cdot dx$$

Let $1 + x^2 = t \text{ or } 2x.$

$$\frac{dx}{dt} = 1 \text{ or } 2x. dx = dt$$

$$\int_0^1 \frac{2x}{1+x^2} \cdot dx = \int_0^1 \frac{1}{t} \cdot dt$$

$$= [\log_e |t| + c]_0^1$$

$$= \log_e |1| - \log_e |0|$$

The solution, as stated above, is incorrect as while substituting $x^2 + 1$ as t; the new values of limits for the variable t have not been replaced.

The correct solution will be as under:

$$\int_0^1 \frac{2x}{1+x^2} \, dx$$

Let $1 + x^2 = t$ or $2x \cdot \frac{dx}{dt} = 1$ or $2x \, dx = dt$

When x = 0, t = 1 and when x = 1, t = 2

$$\int_0^1 \frac{2x}{1+x^2} \, dx = \int_1^2 \frac{1}{t} \, dt = [\log_e |t| + c\,]_1^2$$

$$= \log_e |2| - \log_e |1| = \log_e |2| - 0 = \log_e |2|$$

## xv. Confusion over $e^x$ and $a^x$

Here $a$ and $e$ both are constants, and that is why the confusion exists. The constant e is a different constant given by,

**$e \approx 2.71828.....$**

The following series is used to calculate the value of e:

$$e^x = 1 + x + \frac{x^2}{2!} + \frac{x^3}{3!} + \frac{x^4}{4!} + \frac{x^5}{5!} + \cdots$$

$$\frac{d}{dx} e^x$$

$$= \frac{d}{dx}\left(1 + x + \frac{x^2}{2!} + \frac{x^3}{3!} + \frac{x^4}{4!} + \frac{x^5}{5!} + \cdots\right)$$

$$= \left(0 + 1 + \frac{2x}{2!} + \frac{3x^2}{3!} + \frac{4x^3}{4!} + \frac{5x^4}{5!} + \cdots\right)$$

$$= 1 + x + \frac{x^2}{2!} + \frac{x^3}{3!} + \frac{x^4}{4!} + \frac{x^5}{5!} + \cdots = e^x$$

Here a is any constant, and it may include e as well:

$\frac{d}{dx}a^x = a^x . \log_e a,$

In the above formula if a = e,

$$\frac{d}{dx}(e^x) = e^x . \log_e e = e^x . 1 = e^x$$

Thus, the formula $\frac{d}{dx}e^x = e^x$ is a special case of the formula $\frac{d}{dx}a^x = a^x . \log_e a$.

## xvi. *Double derivative of parametric functions*

For a parametric function, x = f (t) and y = f (t), derivative of y w.r.t. x is

$$\frac{dy}{dx} = \frac{dy/dt}{dx/dt}$$

However, students commit a mistake by extending the formula for getting a double derivative as follows:

$$\frac{d^2y}{dx^2} = \frac{d^2y/dt^2}{d^2x/dt^2}$$

**Example:**

If $y = t^2$ and $x = t^3$. Find $\frac{d^2y}{dx^2}$

**Incorrect Solution:**

$dy/dt = 2t$ and

$$\frac{d^2y}{dt^2} = 2$$

$dx/dt = 3t^2$ and

$$\frac{d^2x}{dt^2} = 6t$$

Therefore,

$$\frac{d^2y}{dx^2} = \frac{d^2y/dt^2}{d^2x/dt^2} = \frac{2}{6t} = \frac{1}{3t}$$

**Correct Solution:**

$dy/dt = 2t$ and $dx/dt = 3t^2$

Therefore,

$$\frac{dy}{dx} = \frac{dy/dt}{dx/dt} = \frac{2t}{3t^2} = \frac{2}{3t}$$

Now,

$$\frac{d^2y}{dx^2} = \frac{d}{dx}\left(\frac{dy}{dx}\right) = \frac{d}{dx}\left(\frac{2}{3t}\right) = \frac{d}{dt}\left(\frac{2}{3t}\right) \cdot \frac{dt}{dx}$$

$$= \frac{\frac{d}{dt}\left(\frac{2}{3t}\right)}{\frac{dx}{dt}} = \frac{-\frac{2}{3t^2}}{3t^2} = \frac{-2}{9t^4}$$

## *xvii. Confusion Over Relation and a Function*

The definition of the **Cartesian Product** of two sets A and B is as under:

A x B = $\{(x, y): x \in A$ and $y \in B\}$

A set R is a **Relation** from A to B if R $\subset$ A X B ( R is a subset of A X B).

Thus, if A = $\{1, 2\}$ and B = $\{3, 4\}$

A x B = $\{(1, 3), (1, 4), (2, 3), (2, 4)\}$

R = $\{(1, 3), (2, 3)\} \subset$ A X B and hence is a **Relation**.

A **Function** is a **Special Relation** in which repetition of first elements of the ordered pairs never takes place.

$R_1$ = {(1, 3), (2, 3)} ⊂ A X B and hence it is a Relation, and its first element is unique, and hence it is a Function also. However, $R_2$ = {(1, 3), (1, 4)} ⊂ A X B and hence is a Relation but its first element one has appeared twice, and hence it is not a Function.

Students should note that all Functions are Relations but all Relations are not Functions.

## xviii. *Wrong Meaning of Inverse of a Function*

In Algebra, students may write $x^{-1}$ or reciprocal of x as 1/x, which is perfectly fine.

Students try to imitate it while writing the inverse of a function as:

$f^{-1}(x) = 1/f(x)$, which is not correct.

If $f(x) = \sin x$, $f^{-1}(x) = \sin^{-1} x \neq 1/\sin x$

*xix. $f(x+y) = f(x) + f(y)$*

Students often assume that in notation of function f (x) stands for f. x (f multiply by x) and thus apply distributive law on functions incorrectly as follows:

f (x + y) = f(x) + f (y), which is not true.

Thus if f(x) = log x

Then, f (x + y) = log (x + y) ≠ log x + log y

Similarly, if f(x) = $x^2$

Then, f (x + y) = $(x + y)^2$ ≠ $x^2 + y^2$

*xx. $f(c.x) = c.f(x)$*

$f(c.x) = c.f(x)$ is another wrong notion.

Thus if $f(x)$ = log $x$

Then, $f(cx)$ = log (c$x$) ≠ c.log $x$

Similarly, if f($x$) = $x^2$

Then, $f(cx)$ = $(cx)^2$ ≠ c.$x^2$

# A WORD OF THANKS

Thanks for reading this book!

If the readers enjoyed the book or found it useful, they may kindly post a short review on Amazon via [Link](Link).

Their support will make a difference to for making this book even better.

Thanks to the readers again for their support.

RAJESH SARSWAT

Made in the USA
Middletown, DE
16 August 2023